all about rats
by howard hirschhorn

Distributed in the U.S.A. by T.F.H. Publications, Inc., 211 West Sylvania Avenue, P.O. Box 27, Neptune City, N.J. 07753; in England by T.F.H. (Gt. Britain) Ltd., 13 Nutley Lane, Reigate, Surrey; in Canada to the book store and library trade by Clarke, Irwin & Company, Clarwin House, 791 St. Clair Avenue West, Toronto 10, Ontario; in Canada to the pet trade by Rolf C. Hagen Ltd., 3225 Sartelon Street, Montreal 382, Quebec; in Southeast Asia by Y.W. Ong, 9 Lorong 36 Geylang, Singapore 14; in Australia and the south Pacific by Pet Imports Pty. Ltd., P.O. Box 149, Brookvale 2100, N.S.W., Australia. Published by T.F.H. Publications Inc. Ltd., The British Crown Colony of Hong Kong.

Cover photograph of a pet rat by Dr.
Herbert R. Axelrod. All photos, not
otherwise credited, by Marvin
Apfelbaum.

ISBN #0-87666-217-3

Table of Contents

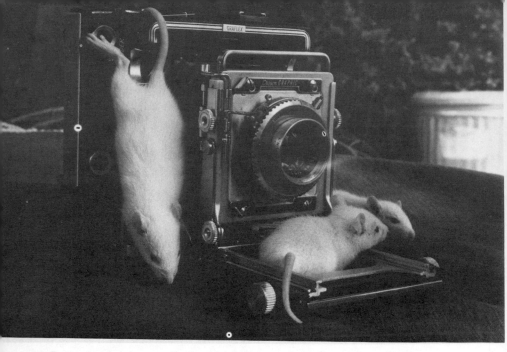

Pet rats are by-products of laboratory inbred strains. They are available in interesting color patterns (see facing page) and as white albinos (above). Rats are intelligent, clean pets. They are easily tamed and trained.

INTRODUCTION

Your pet rat is a mammal—a furry, warm-blooded creature which suckles its young. It lives fast, breathing 2,000 times a minute and maintaining a body temperature of about 100° F. or even lower (varying in different kinds of rats). Contrary to popular opinion, not all rats are destructive or undesirable in human company. Some rats are of immense help to farmers. Farmers distinguish between rodent friends and rodent enemies. Most rats fit nicely into the ecological balance of nature. Even some of the "destroyer" rats have become excellent pets for the right master. Rats are intelligent and attractive as pets, and the domestic laboratory rat has made and continues

Interesting mazes and puzzles keep pet rats busy figuring out how they can get their food rewards. Various mazes are shown on these two pages.

to make some compensation for its wild cousins on destructive bents. Rats are used in psychological research, especially in experiments on learning (such as going through mazes), in medical tests to diagnose human illness, and in pharmacological, toxicological and biochemical laboratories engaged in innumerable research projects.

A world of variety not only for the tender loving company they can provide for a child and adult alike who want to keep them for pets, but also for the naturalist who wishes a microworld of life in his own home where he can fully observe and photograph at his leisure the activities of these intriguing animals who are sometimes inaccessible in nature because of their nocturnal and other-

Rats can be fun!

wise obscure habits. Generation after generation of rats can be born and grown in one's own home.

Small mammals are ideal pets for several reasons: (1) They are fun and bring happiness to many a person who craves an intelligent pet but cannot keep a farm-size menagerie. (2) They provide an insight into nature at a microworld level which the hobbyist can control, generation after generation, breeding just the kind of pets he or she desires.

Some wild animals — when first caught — fight captivity with tooth and nail, or perhaps passively just pine away. Most animals, however, enjoy and can recognize freedom from the constant fear and gnawing hunger of the wild state when they are appreciated as pets.

Wild species are, therefore, described in this book because a good pet book should cover the eventuality of the pet lover bringing home an appropriate animal from the wild.

A reputable pet dealer is a preferred source of pet rats, however. This is important because wild rats can be

Even girls like rats as pets. Photo by Louise Van der Meid.

vicious and/or carry dangerous diseases. However, because animal lovers sooner or later come across a furry creature in the wild and attempt to tame it, and because some of the techniques of taming apply to young domestic animals as well as to wild ones, this book discusses what to keep in mind when handling wild specimens. **Do not attempt to handle wild rats** unless you are certain of their kind and health. Feed them through the cage if you must, but be cautious with any wild creature.

Let us see just where the rat fits into the scheme of nature. Rats are rodents. So are mice, and mice may look and act very much like rats. Dividing lines between rat and mouse are not always readily apparent, some rats and mice belonging to the same family. Generally, a rat is larger and has a longer, more pointed snout.

And then there is the confusion about rabbits and hares belonging or not belonging to the rodents. They do not. They used to be classified along with rats and mice,

A baby rat is compared to a baby mouse of the same age. Rats are larger than mice. Photo by Louise Van der Meid.

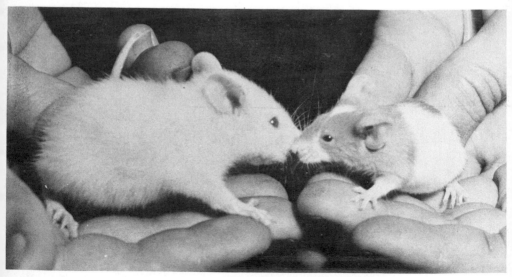

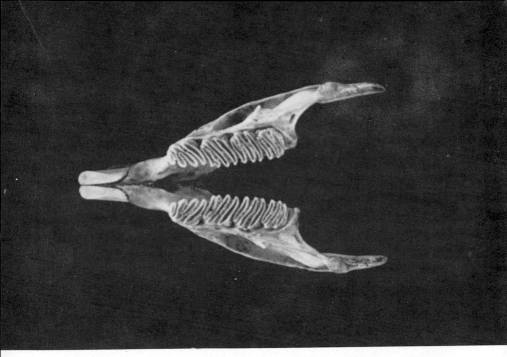

The jawbones of a climbing or tree rat. Rats are rodents and some mice and rats belong to the same family. Photo courtesy of the American Museum of Natural History.

but are now considered separately as *lagomorphs*. Rabbits and hares, along with pikas, belong to the lagomorphs. In general, long ears, long hind legs and a short, cotton-puff tail characterize rabbits and hares. They characteristically hop and leap (but the marsh rabbit may occasionally walk or run in a doglike manner) and are vegetarians. Rodents have two incisors in the upper jaw, but lagomorphs have four incisors in the upper jaw. Lagomorph teeth are made for constant munching. Lagomorphs are herbivorous to the point of being serious agricultural pests when too many are at large. It may appear that a lagomorph has only two incisors just like the rodents, but if one looks carefully, one can see two large incisors in the rabbit's (or hare's) upper jaw, and two additional smaller incisors behind those front upper incisors.

11

Such small furry animals have, historically, been pets and laboratory animals as the hamster above, the guinea pig below, the various colored house mice on top of the facing page and the new long-haired hamster on the bottom of the facing page. Photos by Mervin F. Roberts.

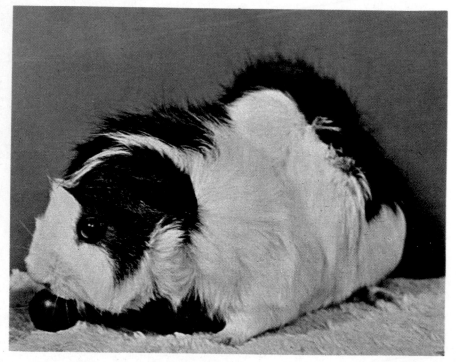

The Rodents or Gnawers

The number of species in the rodent order exceeds the number of species in many other mammalian groups. At least three hundred forty-five families of rodents account for 6,400 species, subspecies or races. A common trait of the rodent is the presence of an upper and a lower pair of chisel-edged, curved incisors. The incisors grow all the time, their length being controlled by the animal's constant gnawing as well as the abrasive action of upper and lower teeth against each other. In addition to the sharpening action of incisor against incisor, this opposing action preserves the life of the rodent: if one incisor is lost or poorly positioned in the animal's mouth, the opposite incisor continues to grow, gradually arcing into the mouth until it pierces the roof of the mouth and thence into the brain. Or it may simply fill the mouth and keep the animal from eating, thus dooming it to a death by starvation. The incisors are separated from the rest of the mouth by a lipfold which protects a rat from getting wood, metal, or any other substance into its mouth; that is, the incisors are used as a scraping tool—a plane—but do not necessarily act as chewing tools for the rat.

Vibrissae or tactile hairs, or just plain *whiskers*, are an essential part of animals built for "nosing around." They are generally found growing near the lips, eyes or cheeks of mammals. Even whales—our largest mammals—have hairs about the mouth, almost as if they were mere tokens of the whale still being a mammal, like man, despite its aquatic habitat. Cats, notably, rely on tactile hair to make their pussyfooting way through obscure and jumbled passages at night. Bats were once thought to completely rely on such hairs to be able to avoid strings stretched across a darkened room, but now we know that the bat's "radar" signals also contribute to their expert flight.

Studies have shown that the rat's vibrissae—when

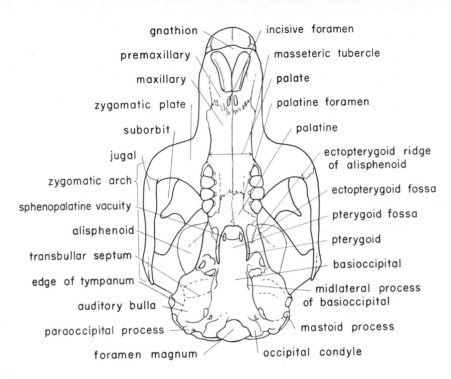

gnathion — incisive foramen
premaxillary — masseteric tubercle
maxillary — palate
zygomatic plate — palatine foramen
suborbit — palatine
jugal — ectopterygoid ridge of alisphenoid
zygomatic arch — ectopterygoid fossa
sphenopalatine vacuity — pterygoid fossa
alisphenoid — pterygoid
transbullar septum — basioccipital
edge of tympanum — midlateral process of basioccipital
auditory bulla — mastoid process
paraoccipital process —
foramen magnum — occipital condyle

A typical rodent's jaw.

the animal is running in the open—extend out beyond the snout, where they evidently play a sort of distance—sensing role. The hairs which are placed most forward sweep the ground ahead of the running rat, the failure of the hair to detect any surface ahead perhaps warning of an impending fall. Side hairs on the snout "scan" edges and vertical surfaces as the rat scurries along. The rat's poor vision is thus in some measure compensated by these tactile hairs. (The cat, too, which also relies extensively on vibrissae, has poor vision as far as immobile objects are concerned, although movement quickly attracts its eye.)

Rodents are usually small, terrestrial animals, but there are burrowing, arboreal and aquatic ones, too. Most are vegetarians, but there are also carnivorous and omnivorous ones.

Rodents are universally distributed, being represented not only by rats and mice themselves, but by squirrels (including flying ones), marmots, beavers, hamsters, gerbils, lemmings, voles (or field mice), the dormouse, pacas, capybaras, goundi, agouti, copypu, rat-moles, porcupines, and guinea pigs.

Rats are available in many beautiful colors (see above and facing page). They are also available in many genetic strains, inbred for many generations, making them ideal for laboratory animals since every strain has its peculiar reaction to various experimental situations. Photos by Dr. Herbert R. Axelrod.

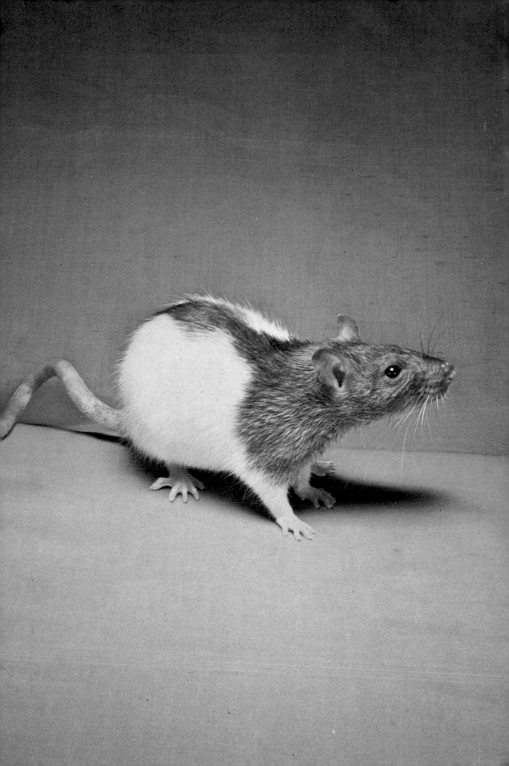

Rats (*and the Mice Scientifically Classified Along With Them*)

Of all the mammals, the best suited to all situations are the rats and mice (that is, of course, except for *Homo sapiens*). New World rats and mice adapt to wide ranges of habitats: open spaces, rocky and mountainous ground, tropical forests, swamps and aquatic or riverine areas. In short, rats live anywhere. This fact is realistically illustrated by a member of an Arctic expedition (E.K. Kane on the nineteenth century Grinnell Expedition):

> *We have moved everything out upon the ice, and, besides our dividing moss wall between our sanctum and the forecastle, we have built up a rude barrier of our iron sheathing to prevent these abominable rats from gnawing through. It is all in vain. They are everywhere already, under the stove, in the steward's lockers, in our cushions, about our beds. If I was asked what, after darkness and cold and scurvy, are the three besetting curses of our Arctic sojourn, I should say, Rats, Rats, Rats. A mother-rat bit my finger to the bone, as I was intruding my hand into a bearskin mitten which she had chosen as a homestead for her little family. I withdrew it of course with instinctive courtesy; but among them they carried off the mitten. . .*

A coypu   rat. A.M.N.H. photo.

The following list of rats, mice and "closely allied creatures" will give some idea of the diversity of the rat and mouse families.

I. **Heteromyidae** family (cheekpouch family): pocket mice, kangaroo mice, kangaroo rats.
II. **Cricetidae** family: muskrats, lemmings, voles, harvest mice, white-footed (or deer) mice, pygmy mice, grasshopper mice, woodrats, cotton rats, rice (or field) rats.
III. **Muridae** (Old World rats and mice) family: black rats, Norway rats, house mice.
IV. **Zapodidae** (jumping mice): jumping mice.

The Heteromyidae have the following family characteristics: they burrow to build nests, are nocturnal, inhabit arid or semi-arid habitats (and do not need drink-

The black rat, *Rattus rattus*. A.M.N.H. photo.

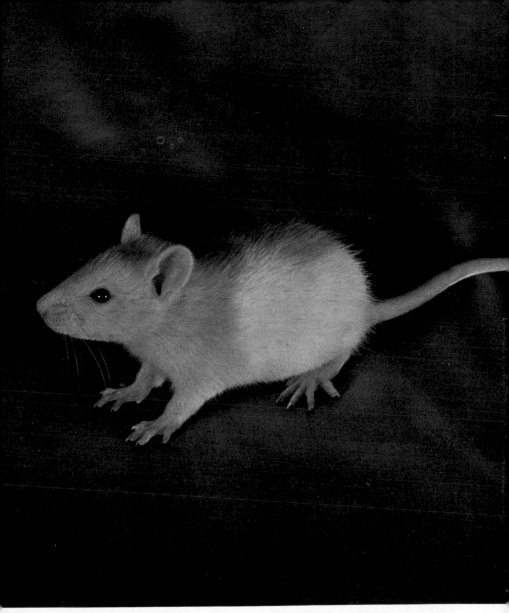

Pet rats are best kept isolated and individually cared for, as the cream colored rat above. On the facing page we see a typical cage filled with rats available at rat wholesalers. Photos by Dr. Herbert R. Axelrod.

Pack rats in a simulated natural habitat. Photo courtesy of the American Museum of Natural History.

A young Branick's rat. Photo courtesy of the A.M.N.H.

ing water), are small, have fur-lined cheekpouches and have weak forepaws but strong hind limbs; their tail is usually longer or equal to their body length; the front faces of the upper incisors are grooved (except for the Mexican pocket mouse); they are not destructive to man's activity (they are indeed beneficial in some cases because they eat weed seeds).

The Cricetidae family has the following characteristics: small to moderate in size (two to ten-inch body length plus two-fifths to eight inches of tail, except for the fourteen-inch body of the muskrat); its members have five-toed hind feet (usually four-toed front feet but a few have five-toed front feet), long tails (except voles and lemmings), large eyes and ears (except voles and lemmings), varied habitats (aquatic, arboreal, rocks terrestrial).

The Muridae are colored a dull gray-brown to black. Long, hairless tails of uniform color are characteristic. The white laboratory mouse and rat are albino forms of the house mouse and the Norway or brown rat, respec-

tively, and all belong to this family. The Muridae represent almost a fourth of all four-footed creatures in North America. Wild forms of the Muridae are usually destructive. Laboratory specimens, however, are a valuable aid to mankind. As pets, some are satisfactory if kept properly. (Do not permit these albino forms of the wild Norway rat and house mouse to escape, for they will breed right back into the wild pest population!)

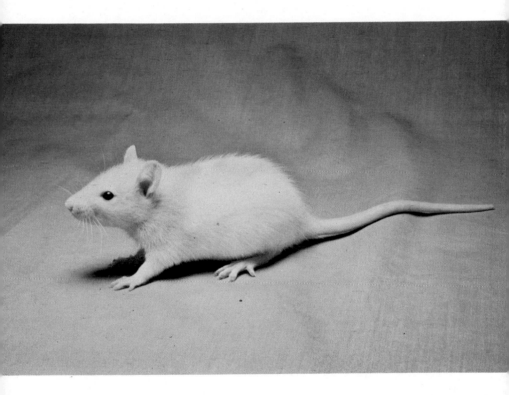

Pet rats must be tamed and constantly handled (see facing page) in order for them to remain tame. A rat must be securely held and the best way is to grasp the rat at the base of his tail while your other hand is wrapped lightly around his body. The buff-colored rat above poses very naturally because he is tame and unafraid of humans.

Members of the Zapodidae family are small to medium in size and have very long tails and large hind feet. The color of the body is yellow-orange on the flanks, darker on the back, and white on the abdomen. Buff or white edging trims comparatively small ears. Grooves run down the front surface of the upper incisors. Winter hibernation and a moist meadow or forest home are characteristic, too.

The following short descriptions of individual rat species are given for a general orientation as to what to expect from wild or domestic pets.

At one time rats were considered pests and bearers of many diseases (photo below, courtesy A.M.N.H.) but now they are friendly pets (top facing page). The capybara, a gnawing mammal from Central and South America (bottom facing page) has not been domesticated. Photo courtesy of the American Museum of Natural History.

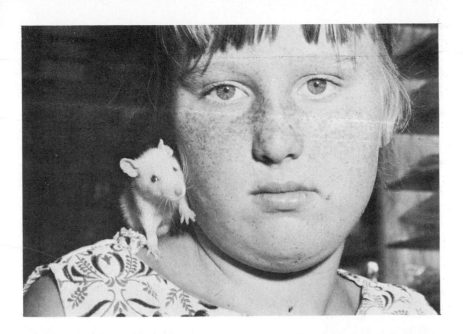

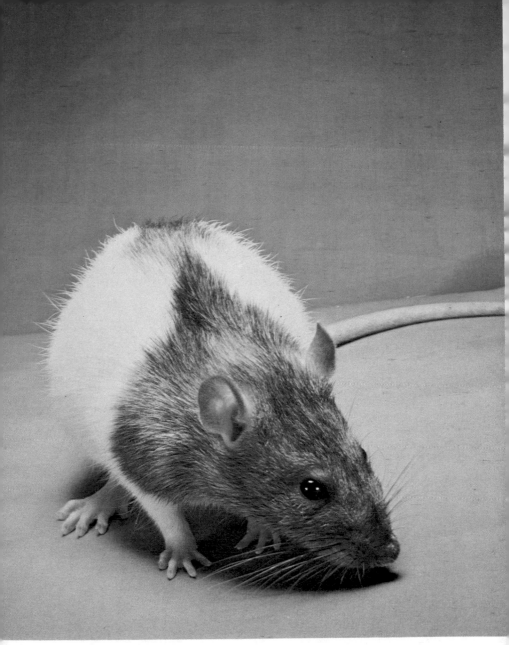

Rats have five toes and long whiskers. They make interesting pets, especially for nighttime people, since rats are nocturnal. Unless you know the rat (and he knows you) don't handle him! Photos by Dr. Herbert R. Axelrod.

The kangaroo rat is used in limited areas as a laboratory animal. Sufficient stocks do not exist for supplies to the pet trade. Photo by the A.M.N.H.

The Kangaroo or Pocket Rat

These are tiny replicas of kangaroos — long hind legs, a long, hairy tail (with a tuft at its end), and short, weak forearms held kangaroo-like in front. Kangaroo rats are especially cute when they try to identify an unfamiliar object near them: they will kick sand at it to see what it does and then form their decision. Clean and mild-mannered, the foot-long species of kangaroo rats (total length, including much tail!) is quite at home wandering about outside of its cage. The kangaroo rat maintains its thrifty nature even in captivity; he will hide nuts or grain in the most unexpected niches about your home. In his native plains and deserts, where he lives as part of a colony, he uses his external, fur-lined cheekpouches to haul loads of food into his burrow.

The Kangaroo rat breathes at a rate of nine hundred fifty breaths a minute and bounds at nine to twelve miles per hour when frightened. And speaking of figures, here is more of the kangaroo rat's track record: it runs five to twelve miles per hour for sixteen to fifty-two feet when closely pursued (California kangaroo rat), and leaps vertically at eight miles per hour.

The abdomen is always white. They have white facial patterns and perhaps black in some individuals. There are usually horizontal white stripes across the thighs, meeting with the tail.

Here is a little life sketch of a tame kangaroo rat who was quite at home in a glass-fronted metal box measuring two by three feet, with the floor strewn with dry leaves. In one corner was a branch, under which the rat apparently felt it possessed some semblance of a burrow, and in which it slept during the day. This rat dusted itself dry in the pan of dry soil (provided in another corner of his box) whenever it accidentally became wet from splashing in the water container. The rat was fed raw carrots, lettuce and sunflower seeds, as well as corn in season. Besides noting this seemingly adequate diet and life style, we

The banner-tailed kangaroo rat. Photo by E.R. Kalmbach, courtesy of the USBSFW.

The pouched rat sometimes called a gopher. Photo courtesy of the American Museum of Natural History.

should also note that although the kangaroo rat is typically a nocturnal creature, this pet gladly accepted fresh food *during daylight hours*. Rats can be adaptable.

The Woodrat

The woodrat (alias *traderat* alias *packrat*, both of these names for a very good reason) should have a spacious and dry cage, plenty of food and respect for its privacy during the first days in its new home, and you will witness the original "Honest Abe" at work. For every (valuable, it seems) item picked up by a traderat on its nightly meanderings, it will deposit in its place a (riduculous, it seems) trinket or piece of junk as if to barter his items for yours.

←

On the facing page, a rat trained to stand and beg for his food. Photo by Dr. Herbert R. Axelrod.

The Merriam kangaroo rat. Photo by E.R. Kalmbach, USBSFW.

A certain M. Chase aptly described this propensity in 1877 (American Journal of Science):

"This house was left uninhabited for two years, and, being at some distance from the little settlement, it was frequently broken into by tramps who sought a shelter for the night. When I entered this house I was astonished to see an immense Rat's nest on the empty stove. On examining this nest, which was about five feet in height, and occupied the whole top of the stove (a large range), I found the outside to be composed entirely of spikes, all laid with symmetry, so as to present the points of the nails outward. In the center of this mass was the nest, composed of finely divided fibers of the hemp packing. Interlaced with the spikes we found the following: About three dozen

knives, forks, and spoons, all the butcher knives, three in number, a large carving knife, fork, and steel, several large plugs of tobacco; the outer casing of a silver watch was disposed in one part of the pile, the glass of the same watch in another, and the works in still another; an old purse containing some silver, matches, and tobacco; nearly all the small tools from the tool closets, among them several large augers. . . all of which must have been transported some distance, as they were originally stored in different parts of the house. The articles of value were, I think, stolen from the men who had broken into the house for temporary lodging. I have preserved a sketch of this iron-clad nest, which I think unique in natural history."

The bushy-tailed woodrat. Photo by V. Bailey, USBSFW.

These rats are clean-mannered, bright and safe; the bushy-tailed species is particularly attractive. Woodrats grow to about ten inches long plus eight inches of tail. They build twig-pile houses in rocky areas, and stick and cactus houses on the plains, or, on the west coast, they nest in oak trees. They are hairy and do not have a scaly tail. The fur is soft. Woodrats are nocturnal and rarely seen. The only mischief they get into is the petty larceny, described above, when a prospector leaves his cabin too long.

The male woodrat's thumping habit (beating on the ground with his feet) attracts females to him quite rapidly. The female woodrat gives birth to one to four (average two) young woodrats thirty to forty-three days after mating during the April-December breeding season. Babies are furred by the fourteenth day after birth and independent by three weeks. Females mature and mate in about six months.

The woodrat. A.M.N.H. photo.

A wild woodrat. A.M.N.H. photo.

Some species of woodrat climb trees and build nests twenty or more feet up in a tree.

Wild woodrats have been observed eating grass, leaves, bark, bulbs, cactus stems, fresh fruit, dry seeds, nuts, mushrooms and succulent cacti.

The Cotton Rat or Marsh Rat

The cotton rat lives in moist regions among tall grasses and weeds. It is a medium size to fairly large rat midway between the size of the house mouse and the Norway or brown rat. Its ears are almost hidden in its fur, its head is somewhat broad and its nose is pointed. Its sparsely haired, scaly tail is shorter than the head and the body.

Coloration of the coat is identical for both male and female cotton rats, with very little seasonal variation. The color is yellow-gray above, with black grizzling. This coloration is somewhat paler on the sides. The underparts are ash to off-white. The feet are gray-white. The tail is

The bushy-tailed woodrat thumps with his tail to attract the female. Photo by N.H. Kent, USBSFW.

black above and gray underneath. The ears are lead gray at their bases. The coloration of young cotton rats is somewhat like the adults but without the grizzling.

Cotton rats have no definite mating season and pair when one month to six weeks of age. Twenty-seven days later, a litter of four to twelve (four to eight average) young is born. The young cotton rats leave the nest and strike out on their own when they are five to six days old. Cotton rats have their incisors even before they are born.

Take care with these rats; they may be vicious. Stronger cotton rats may kill their weaker cagemates. Much of the cotton rat's violent nature, however, has been bred out—the cotton rat has been domesticated (that is, bred in captivity) long enough for it to have become of interest as a pet as well as a laboratory species.

In nature, cotton rats construct runways through grass and other plants in the field. They eat the stems of these plants, leaves and various kinds of seeds, as well as green and ripening grains where they may be found. In addition, cotton is pulled from the bolls by the cotton rat, and the seeds eaten from the cotton back at the burrow.

The muskrat has a flat tail and webbed hind feet. It always is found near water and is rarely tamed. Don't try to get adults as pets; perhaps very young muskrats might be tamed. Photo courtesy of the American Museum of Natural History.

The Muskrat or Musquash Rat

Muskrat fur is known in the trade as Hudson seal, French seal, red seal, electric seal, river mink, and Russian otter. This rodent's most outstanding traits are its long, naked, flattened tail, compressed from side to side, and its musk-like odor. A head and body length of ten to fifteen inches plus a ten-inch tail gives the muskrat a rather large appearance. Its ears are barely visible in the long fur. The coat is dense and soft underneath, and the outer part consists of long, hard, glossy hairs. The muskrat has been called a specialized, and larger, version of the vole or meadow mouse.

The muskrat—its hind feet partly webbed and larger than its forefeet—is built for aquatic life. It mates in the water, staying submerged for as long as twelve minutes. A litter of two to eleven (five to six average) blind, naked young is born nineteen to forty-two days (twenty-two to thirty average) after mating. Young muskrats wean in thirty days and are then expelled from the den by mother muskrat.

The muskrats, who live in open water and along the margins of bodies of water, build conical nests of marsh plants two to three feet above shallow water, or they may burrow into the banks. In northern climates it mates from April to August, but in southern climates it mates during the winter.

Although a single muskrat in your garden pool might be a fine pet for you, some people do not appreciate their destructive tendencies in ornamental ponds. Muskrats which lived in the Bronx River when bogs in the New York Zoological Park were being converted to ponds, moved right into the newly created park ponds. And muskrats do tear up lily bulbs and other water plants in ornamental ponds. If muskrats are crowded in captivity—and that means more than about a dozen in an enclosed pond thirty feet square—they then may turn vicious and perhaps cannibalistic.

Natural pond enclosures for muskrats can be constructed of concrete barriers three and a half feet high, going down three to four feet underground to prevent the muskrat's digging its way out. Provide thick-stemmed plants such as reeds, and the muskrat will probably construct its own house in a shallow part of the pond.

Muskrats in nature eat shellfish, frogs, reptiles, young birds, dead muskrats, vegetables, insects, plant stems, roots, leaves and fruit. A muskrat will dive into a body of water, come up with a freshwater clam or mussel, open and swallow the contents and then, if still hungry, dig out a root with a slash of its forelegs and eat that, too.

The sand rat from South America and Africa. Photo courtesy of the American Museum of Natural History.

Captive muskrats have been fed rolled and whole oats, corn, apples, lettuce, wheat and rice. Another diet for captives consists of corn, marsh plants, raw or boiled crabs and fish. Still another diet, a zoo one, is made up of crushed oats, carrots, celery, green vegetables, apples, bread and dry or canned dog food (sometimes fortified with cod-liver oil).

Rice Rat or Swamp Rice Rat

Another large rat is the rice rat, which is about a foot long. This rat has also been called a mouse (marsh mouse, rice field mouse), but it has a stouter body than the house mouse. Its nose is pointed, and its ears are almost covered by the fur. There is not much hair on the tail, and the soles of the feet are also hairless.

The male and female rice rats have identical coloration, with some seasonal variation. They are brown above, with buff flanks. Underneath they are gray, with grayish hairs tipped with white. Gray underfur shows through. The top of the tail is dark and the underpart is paler. The feet are white. Young rice rats are slate gray.

Rice rats near the Mississippi Delta, for example, mate February to November, giving birth twenty-five to twenty-seven days later to a litter of one to seven blind baby rats. In six days they can see and in eleven to nineteen days they are on their own.

A Norway rat. Photo by the courtesy of the American Museum of Natural History.

The Norway Rat or the Brown Rat

This agile and courageous rat replaced the black rat in 1775 or 1776 as *the* American pest. The Norway, brown, gray, barn, wharf of house rat is a numerous, despicable creature when wild and engaging in its disease-spreading activities. Its survival fitness is outstanding: it swims well, staying underwater longer than three min-

utes; it can tolerate environments at temperatures of -40°
F to 104° F. It thrives in the city or in the country, and
eats literally everything (leather, paint, books, fabrics,
etc.). Stout boards and lead gas pipes can be gnawed
through by Norway rats, the latter case having once
caused a gas leak which in turn led to a serious explosion.
Gnawed insulation (from electrical wires) has occasioned
fires. Norway rats cooperate with one another in hunting.
Wild ones are known to have attacked pigs, calves and
poultry, as well as human beings. They carry disease (the
plague, for example).

These rats have no definite breeding season (unless
very cold weather sets in), and begin doing so at an age of
seventy to eighty days. A litter of four to twenty (nine to
eleven average) blind, deaf, naked young are born
twenty-one to thirty days after mating has occurred. Ears
open by the third day and eyes open by fourteen to seven-
teen days after birth. These rats wean in three weeks. In
sixty days they are mature enough to breed, and stop doing so after eighteen months of life. Norway rats attain a
length of seven to ten inches plus another five to eight inches of tail. Their respiration rate is 2,000 breaths per
minute and their body temperature ranges from 90° to
101° F.

This rat is the common one seen about cities and ur-
ban areas. Its pointed nose is naked at the end. Its eyes
and ears are large compared with some other rats. Its tail
is somewhat characteristic: there are rings of overlapping
scales on the nearly naked skin.

The male and female Norway rats are identical in
coloration, with some slight seasonal variation. The fur is
coarse. Coloration above is gray-brown and dirty white to
gray below. Black hairs, particularly along the back,
characterize the upper part of the Norway rat. Its feet are
a whitish gray. The tail is dusky above and somewhat
lighter underneath. Young Norway rats are grayer than
the adults.

Rats and mice (these are mice) look so much alike they are hard to distinguish (except for size) by many fanciers. The black and white Dutch mouse is well colored but the slate and white next to it is mismarked. Photo by Harry V. Lacey.

The Black Rat

The black rat was the ordinary house rat in the American Colonies before the Revolution; the Norway rat has since replaced it. Ironically, however, in losing its battle for existence in the wild with the brown or Norway rat, the black rat has survived as *the* laboratory white rat. Gentle albino and colored strains (black-hooded, red or auburn, red-hooded) of the black rat now serve medical research and can be obtained from pet dealers. These rapid breeders are rarely vicious if handled when they are youngsters. Hoary old males even like to be scratched on their heads.

This glossy-haired rat climbs trees, pillages bird's nests and twitters like mice; Norway rats, on the other hand, squeal and do not twitter.

Black rats mate at eleven weeks of age and start heat (for nine to twenty hours) between the hours of 4:00 and 10:00 p.m. Litters are born twenty-two days after mating has occurred. The lifespan of these rats is up to about four years.

Other Rats

Other rats which the avid pet keeper might like to investigate are the South American fishing rat, which lives along mountain streams; the bamboo or root rat of Malaysia, Indonesia, Tibet, India and China; and the tree rat, which is an expert climber. And there is the wading rat (also called tree mouse), shrew rat of the Philippines, cane rat, cloud or rind rat of Asia and the Pacific, Australian water rat, mole rat of Eastern Europe and the

A fully grown hooded rat is larger than your hand.

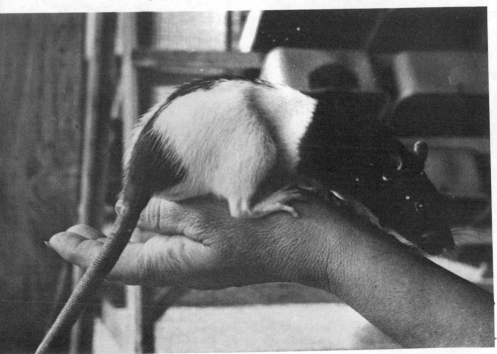

A cane rat from San Antonio, Texas. Photo courtesy of the American Museum of Natural History.

Middle East, and the African blesmols (also called strand rat or sand rat).

Many names, however, sometimes refer to one and the same "basic rat," as, for example, the following rice rats; this listing of "variants" gives an idea of the geographical distribution of this one rat—the rice rat—as well as illustrates how many varieties can occur. Here are some varieties of rice rat: Coues' rice rat, Baja California rice rat, Thomas' rice rat, Azuero rice rat, Cozumel Island rice rat, Jamaican rice rat, Nelson's rice rat, Gatun rice rat, Panama rice rat, Alfaro's rice rat, black-eared rice rat, Talamancan rice rat, Boquete rice rat, Harris' rice rat, Mount Pirri rice rat, St. Vincent rice rat, pygmy rice rat, dusky rice rat, Enders' rice rat, regal rice rat. . . ! But do not confuse varieties with synonyms—some names are precisely the same rat.

The following comparative table will give a notion of the size and proportion of various rats. Figures are averages only, and individuals vary quite considerably, even within a single species.

A jungle rat from the Congo jungle in Zaire. Photo courtesy of the A.M.N.H.

Lengths in Inches

Common Names	Total	Tail	Hind Foot	Ear
Norway rat, house rat brown rat, gray rat, barn rat, wharf rat	15.7	7.5	1.6	— —
Trade rat, packrat brush rat, eastern woodrat	15.5	7.5	1	0.9
Brush-tailed woodrat	15	7	1.7	— —
Cotton rat, marsh rat	10	4	1.3	0.8
Rice rat, rice field mouse, marsh mouse	8.8	4.4	1	0.6
Muskrat, musquash	21	10	3.5	— —
Round-tailed muskrat	13	5	1.5	— —
Pocket rat, kangaroo rat	11.3	7.1	1.6	0.5
Giant pouched rat	30	— —	— —	— —
Giant or cloud rat	15-18	— —	— —	— —

HANDLING YOUR RAT

Everyone says "don't pick up your rat by its tail!" Good advice — the tail might skin off if you grab a fleeing rat by its tail. However, there are times when that is all you can seize. . . but let it go as soon as possible to release the strain on that organ (and it is an organ of balance).

Rats are delicate, small animals and they are easily injured if you grab them and squeeze. Hold them gently and have your petshop manager show you the proper way to hold a rat.

Remember that pet rats usually bite from fear, not anger or vexation. Regular handling makes your new pet docile. To train your pet to sit carefree in your hand, set it on your palm, the forefinger and thumb of your other hand gently holding the base of the tail.

Handle rats as you clean their cage; this helps to tame them. Rats which are frequently handled grow better and faster than neglected rats. Play with them at least two to three times weekly. Animals are *kept* tame; they

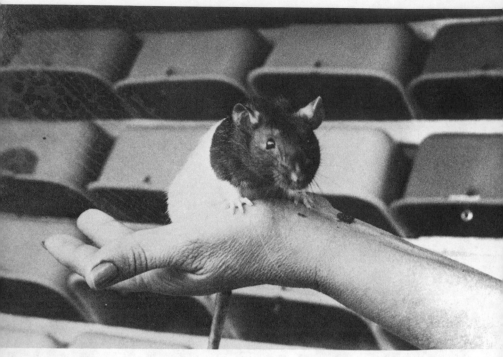

Hooded rats are very popular as pets. . . and they are very easily tamed and trained. The baby albino rats on the facing page have a lot of growing to do. When full grown they will be bigger than your hand.

may revert instinctively to untamed behavior. Let the rat know who is master. Be kind, yet firm and do not show your fear, if you have any. Rats catch on quickly. Associate feeding time with handling and petting. Your rat will then look forward to eating and to your scratching its head while it eats. You can make the same noise (whistle, tch-tch, clicking tongue, kissing sound, raspberry, etc.) each time you feed your rat, and your pet will soon come to associate the sound with you and with food. Hand-feed rather than leave food in the cage except when you are away overnight or even several hours (remember, rodents keep on eating, so leave enough food when you are away).

Fear quickly spreads among all rats in a cage. Wild rats panic easier than domestic rats. Make no sudden movement, or even any slow movement which the rat can instinctively interpret as the approach of a predator.

Again, it is worthwhile to repeat: it is better to obtain rats from reputable dealers. *Wild stock can be vicious or dangerously diseased!*

HOUSING

Metal cages are usually used because rats can gnaw through quite a few substances (they are master rodents!). Take care that metal cages do not become too hot (near radiators) or too cold (in winter, outside, or placed too close to a window). Wooden cages may be suitable if reinforced with heavy-gauge wire mesh, that is, half-inch wire netting.

As a guide in providing adequate room for your rats, allow one hundred forty square inches (14" x 10" x 10") for a brown or Norway doe with her babies. Allow one hundred sixty-eight square inches (14" x 12" x 10") for a breeding pair of brown rats.

More specific recommendations have been drawn up by various authorities as follows:

The photographs on these two pages show ways of getting attention in your school! Some girls like to have the young rats climb on their hair. This is always good for screams and laughter and a sure way to get sent home from school! Modern schoolteachers encourage their schoolchildren to love and understand pets, providing they know how to handle and care for them!

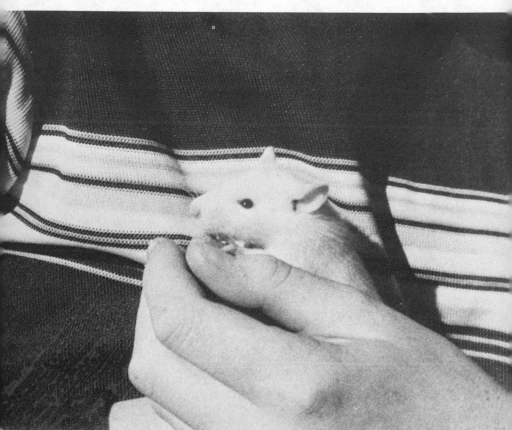

Weight (grams)	Minimum space (square inches/rat)	Maximum rats per cage
Up to 50	15	50
50-100	17	50
100-150	19	40
150-200	23	40
200-300	29	30
Over 300	40	25

Aquarium tanks are fine if they are large enough; they permit an open view, are easy to clean, and are escape-proof (over ten inches high or fitted with a top which still allows air to enter). A layer of soil can be put in these waterproof aquarium tanks, too, thus making your rat's home more attractive than a clinical cage would be.

If a metal dollhouse becomes your rat's home, be certain to ventilate it adequately and provide for a moveable side, front or back, so that the inside can be cleaned properly.

Softwood shavings—especially cedar wood—make an excellent bedding and nesting substance. Sawdust, too, is very good, but it must be made from white softwood; hardwood sawdust may contain harmful natural chemical substances (phenols). Obtain the shavings or sawdust directly from the sawmill or pet shop. Otherwise, it could be contaminated with the droppings or hair of animals. Peat moss, although expensive in some areas, greatly minimizes odor. Its acid content counteracts the decomposition of animal droppings. Shredded paper is an old standby. Renew bedding once weekly whether it seems dirty or not.

Rats naturally huddle and sleep together.

Rats are curious and inquisitive.

Cedar shavings, available at your petshop, make excellent bedding and help keep your pet rat clean.

Once a rat colony hierarchy is established, the admission of new members of the "rat society" precipitates trouble. Scientists have performed an experiment to demonstrate "the fall of a society" of rats. In a rat colony, the strongest rat takes charge over the not so strong ones. A certain balance is struck and community life gets along quite well in the space provided. If, however, a cage with two doorways is made and other rats are admitted simultaneously through these two different doorways, then the leader of the established colony cannot guard them both at once. The social order is endangered. Things begin to go wrong. When a rat starts for food, other rats lie in ambush and attack the food-seeker. Nesting routine is upset and almost all of the baby rats die.

Two types of excellent cages for rats are the aquarium type (above) and the typical hamster cage with exercise wheel (below).

The usual kind of laboratory cage is a plastic tray and a stainless steel top which holds food and water.

Colonies are generally housed in wire-mesh or glass-sided cages with removable wire tops, or else in plastic bins fitted with wire tops, each of which holds an inverted gravity-feed bottle and a hopper to hold pellets or other food. Bottles, boxes, flower pots (on their sides) or metal containers serve as hide-aways or nesting sites.

Demand (or gravity-feed) water bottles, not troughs or pans, are the best way of providing a fresh, clean source of drinking water. Open containers of water get stepped in, overturned, excreted in, and can become an unsightly (if not downright unhygienic) mess.

A gravity-feed water bottle delivers water due to the force of gravity to a perforated cone or drinking tube attached to the neck of the bottle. The animal sucks or licks water from the opening in the cone or tube; air bubbles enter the bottle, replacing the water being drunk. No air enters and no water drops out when the animal does not lick or suck the opening. The bottle will leak if there is a defect or incorrect adjustment in the system. Leakage can be caused by the wrong size hole in the cone or tube, a bad fit between tube or cone and stopper, or stopper and bottle, agitation of the bottle (water bottles cannot be used in moving vehicles), etc. Temperature fluctuations will alternately contract and expand the air in the bottle, driving the water out. Contact between objects and the

A gravity-feed water bottle is introduced to these rats.

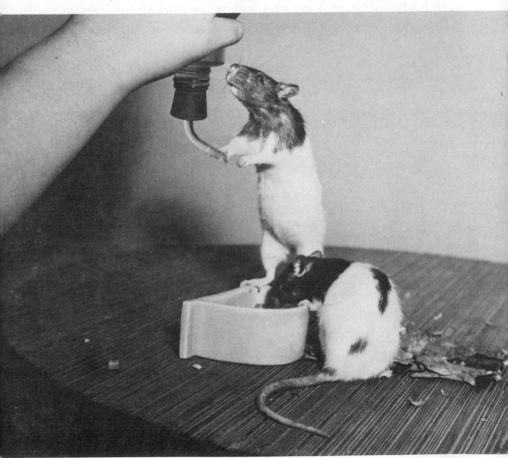

hole delivering the water (or if the animal brushes against it) may start the bottle emptying (the so-called siphoning effect). Fungus, alga, calcareous deposits, corrosion, food, or bedding might block the delivery tube.

Bottles that are too small need to be filled too frequently. Water bottles which are too large may empty themselves by the siphoning effect when they are only half-filled.

When an animal drinks from a bottle, however, it will still be able to foul the water with saliva and the organisms in its mouth, and some organisms multiply rather well in ordinary faucet water. When room temperatures are high, such water may harbor more organisms after a while than when you put it in the cage. So change water often.

Rats' saliva can dirty the water spout, so clean the spout and the water bottle as part of your general cage-cleaning procedure.

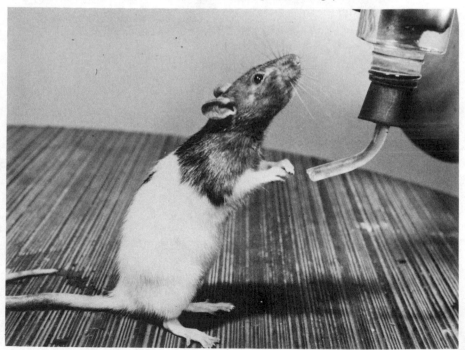

Rats eat as much as they require to satisfy their appetites. They do not overeat. This rat is having a snack in his exercise wheel. The more he exercises, the more he will eat.

RAT FOOD

Rats are easy to feed. Besides feasting on your table remnants, they can also be fed grain, birdseed, whole corn (you will certainly enjoy watching your pet rat hold a corn kernel between its little paws and nibble on it like a squirrel), bread scraps, meat, chick-rearing pellets, seeding grass, cheese (of course!), fat, bacon rind, drippings,

Mice and rats (this is a mouse) eat almost anything, though they prefer grains.

lard, fruit, diced carrots, linseed (for a glossy coat), bones (for keeping the teeth worn down) and hard doggie bones. Rats also eat raw eggs. But not many people knew how the rat got away with the egg:

> *Astute observers noted that the rat pierces the egg-shell and sucks out the liquid contents. Some people, however, propagated the tale that one rat would lie on its back, with the egg on its belly, and have another rat pull it and the egg away like a sleigh.*

Do not feed too much of anything at once; it becomes wet and putrid, forming a fine culture medium for the growth of bacteria and fungus.

RATS CARRYING OFF EGG.

An old time sketch explains how rats DO NOT carry eggs!

You can have your rat trained to eat from your hands.

Rats may be fed in seed cups, but some rats might foul the food if it is open. Test your rat's appetite and taste by offering different types of food and see which he prefers.

Here is a rat food test you may wish to use: present your rat with a series of small jam, herring or peanut-butter jars, each one filled with a different kind of food, and put water in one of them. Each day you will be able to see how much of each kind of food is eaten, and this will be the perfect taste test. The thing to watch for is which one is eaten up in one day with the least waste.

Some breeders provide their colonies of rats with staple diets of canary and sunflower seeds, oats, raw vegetables, and fruit. Cod-liver oil is mixed in with dry or canned dog food. Sometimes meat and fish are added for proteins.

Meat-eating rats need to be kept supplied quite well with fresh meat (insects or whatever the particular species eats); occasionally, substitutes for fresh meat are accepted by the rat. Experimentation will yield much information on how each particular animal is willing (or indeed is even able) to adapt to the food eaten by its human master.

Lactating rats have higher requirements than non-lactating ones; a lactating rat can eat her own weight in dry food in two days.

Young rats will eat bread soaked with milk and warmed to body heat. Hoppers or baskets designed to contain the food pellets are good; rats, however, may climb up onto the food pile, even up on the underside of it, and contaminate it with their droppings or other materials such as saliva. Rats drink about an ounce of water daily. Gravity-operated "demand" water bottles are available which allow the animal to serve itself.

Balanced diet rat pellets are available at pet or feed dealers. These pellets can be placed in self-feed hoppers for making them conveniently available to your pet. Good quality, commercially prepared diets for general feeding purposes, as well as special diets for breeding animals, are available. If you buy commercially prepared rations, do not keep them too long before use. Nutritional quality of stored dried foods drops off with time. Prepared foods, ideally, should be used within four to six weeks after manufacture or compounding.

You should be aware that certain additives (such as estrogen agents) in prepared foods could affect breeding behavior and results.

Rats eat about fifteen grams daily of the ordinary pellet foods on the market. If a good food pellet or cube is

Mature rats eat about half an ounce of food a day.

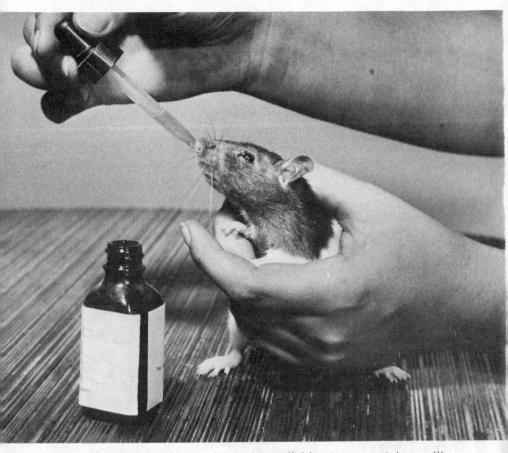

Vitamin and mineral supplements available at your petshop will help maintain your rat's general health and breeding ability. Let them lick it from a dropper if you can.

given, no supplementary food is really needed, except about five grams of cabbage for the cotton rat. "Really need" means for satisfactory growth. Varied supplementary feeding (vegetables, raw liver, cod-liver oil, milk, dried yeast, etc.) will increase growth and breeding capacity to much more than a "satisfactory" level.

Once a daily ration has been decided upon, stick to it, at least for awhile. Abrupt changes may disturb your pets. Enteritis and diarrhea may occur in new arrivals because of such sudden dietary changes. Gradually make changes, if any are to be made, in the diet of new arrivals, if you know what they were receiving before you obtained them to begin with. Do not overfeed newly arrived animals.

For fun and games with feeding, throw a nut into a cage of young rats and watch the hilarious scramble for the tidbit.

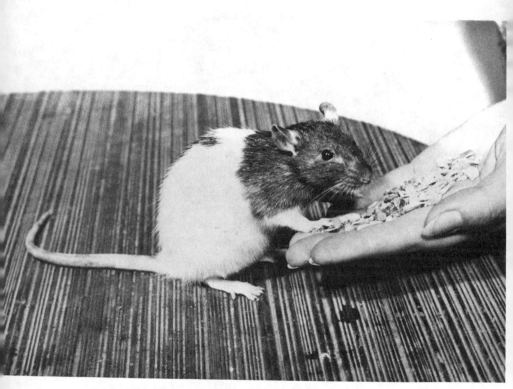

If you keep your pet rat hungry, he will quickly learn to feed from your hand.

The following list shows which foods are definitely known (by observation of wildlife) to be eaten by wild rats, as well as some of the dietary items eaten (and apparently enjoyed) by rats in captivity.

Rat names and synonyms	Diet observed in nature and in captivity
Norway rat, house rat, brown rat, gray rat, barn rat, wharf rat	Omnivorous
Traderat, packrat, brush rat, eastern woodrat	Seeds, grains, leaves, plants, greens, bread, corn, chopped fruits, dog pellets
Brush-tailed woodrat	Seeds, green plants
Cotton rat, marsh rat	Seeds, grasses, green plants, meat, bird eggs
Rice rat, rice field "mouse"	Seeds, grasses, green plants
Muskrat, musquash	Omnivorous; roots and stalks of irises, lilies and reeds; clams, fish, insects
Round-tailed muskrat	Roots, green plants, animal flesh
Pocket rat, kangaroo rat	Herbivorous only: seeds, grain, greens, bread, corn, chopped fruits, dog pellets, carrots
Giant pouched rat	Has been fed dry and canned dog food, grains, green vegetables, fruits
Giant or cloud rat	Has been fed rolled oats, carrots, celery, lettuce, bread, dog food (fortified)

See also discussions of diet under the sections on individual species.

RAT GROWTH CHART (Averages in ounces)

	Small Sherman Albinos Male/Female		Large Sherman Albinos Male/Female		Wistar Albinos Male/Female		Wild Norway Male/Female	
Birth	0.2	0.2	0.21	0.20	0.2	0.18	—	—
1 week	0.47	0.46	0.6	0.57	—	—	—	—
2 weeks	0.88	0.88	1.3	1.2	—	—	—	—
3 weeks	—	—	—	—	1.5	1.4	—	—
4 weeks	2.2	2.0	3.3	2.8	1.8	1.9	—	—
6 weeks	4.3	3.5	6.6	5.2	3.9	3.4	3.0	3.7
8 weeks	6.2	4.6	9.7	6.9	6.0	4.5	6.0	5.4
10 weeks	7.8	5.4	12.0	8.0	7.1	5.2	—	—
12 weeks	8.9	6.0	13.9	8.9	7.9	5.8	8.4	6.8
15 weeks	10.0	6.5	15.5	9.6	8.8	6.3	10.2	8.1
20 weeks	11.5	7.1	17.3	10.7	9.9	7.1	—	—
30 weeks	13.3	8.1	—	11.8	—	—	—	—
35 weeks	—	—	—	—	—	—	15.7	13.3
40 weeks	—	8.5	—	12.6	12.1	8.6	16.2	14.0
45 weeks	—	—	—	—	—	—	16.5	14.6
50 weeks	—	—	—	—	—	—	16.7	15.0
52 weeks	12.8	8.6	—	—	—	—	—	—
57 weeks	—	—	—	—	—	—	16.8	15.3

This chart gives a *general* idea of how weight progresses, depending, of course, on many factors (diet, health, living conditions, season, and so on). Note that females usually weigh less than males. Keep in mind that Sherman and Wistar albinos are laboratory rats, and thus receive balanced diets and other (sometimes) optimal care. Personal pets may be fed inadequately *or* they may be pampered and fed royally. So, again, the above chart is a guide only.

HEALTH

Before buying your rat, check to see whether the dealer's cages are clean. Does the rat have a glossy coat? Are the animal's eyes bright, ears and feet clean? Is the rat plump, but not fat? Does it move freely and uninhibited and not jerk along nervously? Rats are less likely to contract serious disease than most other laboratory animals. Your healthy pet, if fed and housed well, will most likely do well unless it picks up an infection or is injured; or, if its resistance is lowered (due to poor food or environmental conditions) it may manifest one of the typical domestic laboratory rat conditions: salmonellosis, chronic lung condition or a middle ear condition (called labyrinthitis or "circling disease").

Salmonellosis (or paratyphoid fever) is manifested by acute diarrhea, conjunctivitis, and poor general condition. The ray may die in seven to fourteen days following exposure to the causative bacterial organism. Wild rats can infect your domestic rats with it. (Your domestic laboratory rat, incidentally, is still capable of interbreeding with its wild ancestor, the Norwegian or brown rat; therefore, do not let it escape into the wild state.) If the rat does not improve spontaneously, drug treatment is usually of no avail.

Pulmonary or lung conditions cause sniffling. This, too, may or may not clear up spontaneously.

Circling disease (labyrinthitis) is so named because the rat tilts its head, walks in circles, and has difficulty maintaining its balance. This disease is a middle ear infection, and since the ear is where the rat's organ of balance is located (man's, too), an infection affects its ability to walk in a straight line and to keep its balance.

Ringtail should be mentioned here. Believed due to the rat's environment being below fifty percent humidity (in the room where the rat is kept) as well as to cold or

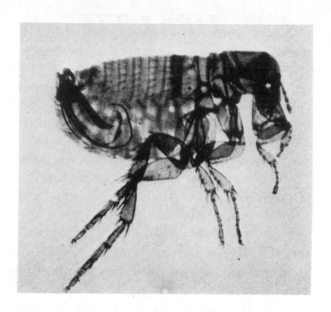

Rats have fleas. These two photos show the rat flea, *Laemopsylla cheopis*. The male flea is shown above and the female flea below. Fleas may carry disease on wild rats, so keep your rat caged and away from wild rats.

draft, it is manifested in the tails and perhaps toes of newly born rats. The tail or toe swells; it may fall off after the formation of a ring of construction around some portion of the toe or tail. Ringtail clears up by itself. Young rats kept in solid bottom cages (and not wire mesh) rarely have this condition.

For mange, ask your dealer for a mite powder or dip. Some skin and fur conditions may be due to vitamin deficiencies rather than to mites or other parasites.

Your petshop will have flea and tick powder for your rat. Hamster, guinea pig, gerbil or mouse powder will also serve your rat's needs.

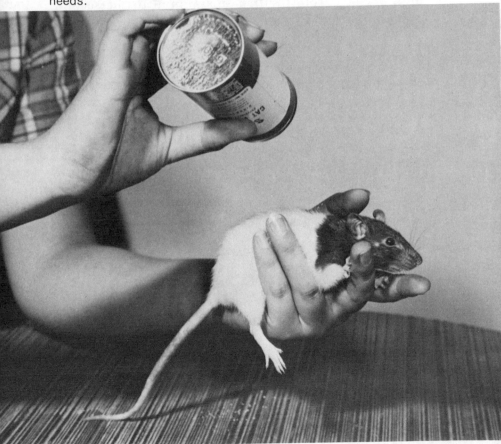

Rats probably obtain their vitamin D by licking themselves; if rats are not allowed to lick their fur, they can develop rickets. The oils and fats in the fur are most likely absorbed through the rat's mouth when it licks, and the vitamin D comes from those oils and fats. This is also true for some other animals (dogs, cats, horses, etc.).

Spread of Disease

The spread of disease among rats in a community housing set-up depends upon the following factors:

1. How many animals are already infected?
2. How infective is the organism? How soon do adequate numbers get from the first sick animal to the second sick animal, and so on?

Sick rats might spread disease to other rats by contaminating the drinking spout with their paws and mouth. Sickly looking rats lose weight, have runny noses and eyes. . . and they look sick!

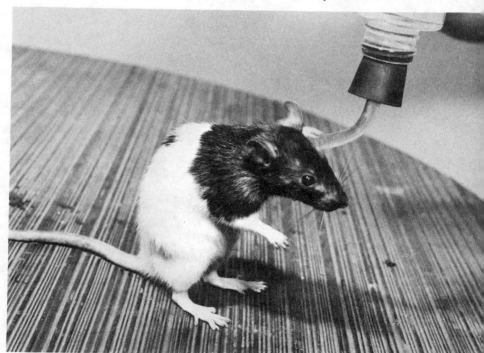

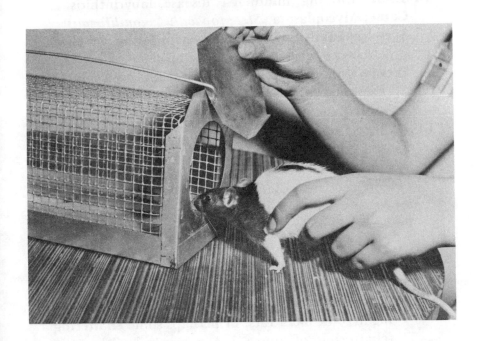

If you suspect a particular rat is sick, isolate him in anything which is safe for the rat and your other furry friends. An old humane-type rat trap is ideal! But be sure you offer food and water while the rat is in isolation.

3. How potent is the disease-causing organism, and how well does it counter the animal's resistance?
4. How resistant is the animal against a particular disease-causing organism? (Some resistance is natural, some is acquired by previous bouts with the disease, and some comes from prior vaccination or inoculation.)

Symptoms, Causes and Control of Diseases

The following table is a brief summary of rat diseases and some of the treatments tried and reported by breeders.

DISEASE: Circling, middle-ear disease, labyrinthitis
  Cause: Mycoplasma, *Streptobacillus moniliformis*
  Symptoms (all not always present): Tilting of head, circling, difficulty in getting up
  Prevention and/or treatment (what has been tried)*: Prevent contact with sick animals. Keep housing clean.

DISEASE: Salmonellosis
  Cause: *Salmonella* sp.
  Symptoms (all not always present): Loss of condition, conjunctivitis, diarrhea. If death occurs, it is in 1 to 2 weeks after infection.
  Prevention and/or treatment (what has been tried)*: Quarantine new arrivals. Disinfect housing.

DISEASE: Epidemic murine pneumonia
  Cause: Virus(?)
  Symptoms (all not always present): Difficult breathing in some older rats, or perhaps some chattering (chattering, of course, is common in healthy rats, too)
  Prevention and/or treatment (what has been tried)*: Newborn rats are infected by females. Strict hygiene will probably prevent appearance of disease (even though many adult rats may harbor the disease organism).

DISEASE: Coccidiosis
  Cause: *Eimeria* sp.
  Symptoms: Diarrhea in weaning rats
  Prevention and/or treatment (what has been tried)*: Provide adequate nutrition to keep rats resistant to effects of infestation. Keep bedding clean.

DISEASE: Parasites on skin and in hair
  Cause: Mites and fleas

---

* Follow instructions provided with all medicines you buy. Seek expert advice. "Doctoring up an animal" can lead, in some cases, to conditions worse than the disease being treated.

Symptoms (all not always present): Skin irritated. Sometimes sores on tail or skin.

Prevention and/or treatment (what has been tried)*: Benzene hexachloride powder, tetraethylthiuram monosulfide (diluted), DDT powder. Repeat treatment once in a week after initial application.

DISEASE: Cysticerosis

Cause: Cat tapeworm

Symptoms: Usually none

Prevention and/or treatment (what has been tried)*: Be wary of sawdust obtained from sawmills; it may contain cat feces which contains the tapeworm.

DISEASE: Ringtail

Cause: Believed to be due to humidity below 50 percent

Symptoms (all not always present): Newborn rats' tails or legs swell, with corrugated-like constriction which may lead to loss of the extremity.

Prevention and/or treatment (what has been tried)*: Increase humidity.

DISEASE: Avitaminosis (vitamin E)

Cause: Diet poor in vitamin E

Symptoms (all not always present): Infertility, resorption of fetus

Prevention and/or treatment (what has been tried)*: Add synthetic vitamin E to diet, unless commercially obtained pellets already contain it.

A Note on Treatment. . . And Not Treating

The treatments suggested here are some of the measures tried by others. Do not consider them sure-cures. Avoid treatment with drugs and medications unless you have expert advice or clearly understand the instructions

* Follow instructions provided with all medicines you buy. Seek expert advice. "Doctoring up an animal" can lead, in some cases, to conditions worse than the disease being treated.

which come with the medication. In many instances, rat breeders "dispose of" an ill rat rather than try to cure it. With pets, of course, one does not wish to take such a drastic step. Consult a veterinarian — a Doctor of Veterinary Medicine, D.V.M. — when in doubt as to the proper, safe treatment.

Hygienic measures go quite far in preventing disease and curing it by allowing the rat's own resistance to help. Nutritional support such as vitamins may help, too. Keep food and water fresh. Keep housing and bedding clean and dry. Remove accumulated droppings. Although ani-

You can use a glass and metal cage to isolate a sick rat, but after the rat is cured or expires, you must sterilize the entire cage by heating it in an oven at 250° for 30 minutes.

If your rat is sick, leave him alone. He needs rest and quiet.

mals should be fondled and played with to keep them tame (and they enjoy playing, too!), let them rest quietly awhile if they seem indisposed. You can still "sit by their bedside" and talk with them until they recover, but reduce any rough-house for awhile.

Terminal Disinfection

Dead rats should be promptly removed from a group of rats, the other rats transferred, and the box cleaned and disinfected. This may stop contagious infections from spreading to healthy rats, although some may already be infected; hygienic maintenance of the cage, however, will hold down the disease manifestations or perhaps limit the severity of any outbreak when and if disease finally appears after an incubation period.

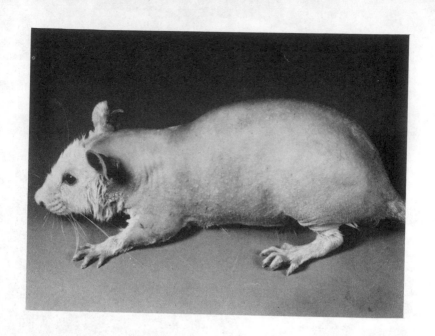

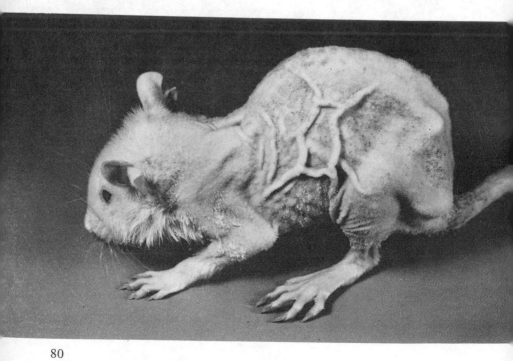

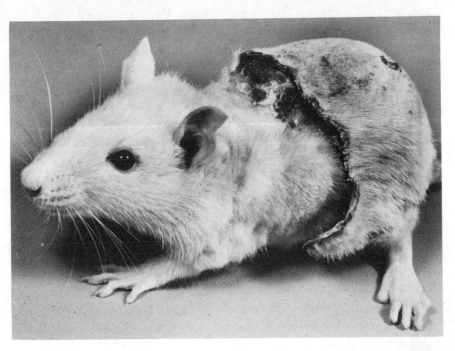

Dr. Hans Selye at the University of Montreal supplied us with this series of three photographs which shows a very rare condition in rats called *calciphylaxis*. It is almost unbelievable but the rat's skin calcifies like a crab and is then shed. This is very rare in mammals, but very common in many insects and crustacea. Rats have as many diseases and disorders as man.

BREEDING

Rats, in general, do not wait for any definite breeding season before they mate. When breeding, the female's genitalia are somewhat swollen, meaning she is possibly in heat. This stage lasts about twelve hours. Definitely swollen genitalia indicates she is in heat and copulation is in full swing. This lasts twelve hours. Then swelling of genitalia begins to abate, and a copulation product is present in the vagina. This lasts fifteen to eighteen hours, then all swelling is gone. Be careful with rats when they are breeding. Some may be vicious. The cotton rat female may be extremely violent to the male and even kill him during mating. Mate such rats early, say at six weeks of age, to avoid this violence.

The pet owner can control to some extent just what offspring are born to his rats. To do this requires some knowledge of genetics—the science of genes and of their selection. Genes, being responsible for the inheritance characteristics or traits (short hair, pink eyes, agouti-colored coat, etc.), can be *selected* for by the owner. That is, he can pick out parents with the desired characteristics and then let these mate. With a little experience, undesirable characteristics can be *bred out* by not letting the animals with those characteristics mate. In general, one can breed rats in three ways: (1) random mating (free-for-all, or colonial system); (2) the harem system (one "pasha" with his bevy of females); (3) the monogamous pair system (nice little "married" pairs kept apart from the others).

The random mating system is a free-for-all, or colonial system. This method does not require the keeping of any exact records and requires only that one take the best-looking males (that is, the most virile-looking one) and put them together with some of the best-looking females of the same age group. This method allows large numbers of animals to be produced and requires hardly any paperwork at all.

The next system or the harem system is where one "pasha" is placed together with his bevy of females. The number of females may range from two to twenty. Some experimentation is necessary to use this system because not all species will mate in such a situation. The harem method requires very little space and the resulting number of rats born is very great. One should be aware, however, that in the harem system newborn rats may be smothered by their older littermates or even by the adults.

In the monogamous pair system one encourages nice little "married" pairs which are kept apart from the others. Individual mating records are possible in this system and this, of course, involves paperwork. The detailed records system is only one disadvantage of this method. Much space, attention and equipment is needed for raising even a small colony.

One should be aware of the existence of linkage when selecting for certain characteristics. Linkage is the

This mother rat is protecting her newborn which are barely visible underneath her body. Don't reach for them or she will very probably attack your hand.

tendency for a group of genes to be inherited together continually from one generation to another. The following section on linkages in rats should not only give an idea of how your pet rat could look, but explain some "peculiar" mannerisms in your pet. Your pet may be that way because of its genetic heritage!

An example of linkage is the albino rat. This rat is a laboratory form of the Norwegian rat which occurs wild. In the laboratory, however, this rat has a white coat and pink eyes. The whiteness and pinkness are both due to lack of pigmentation, the pinkness of the eye being the color of the blood in the capillaries, and not being masked by any pigmentation.

The following genetic glossary gives not only a description of how a rat may look, but also of how it may act. Some of these conditions may go together with others, forming linkages:

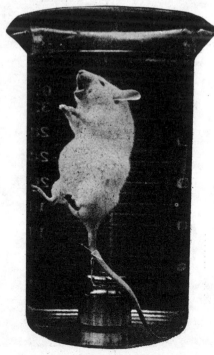

A rat breathing perfluorobutyl-tetrahydrofuran. After an hour's immersion, the rat was inverted to drain the liquid from its lungs, and is now alive and well. (Source: Leland C. Clark, Jr., University of Cincinnati College of Medicine)

Agouti—an irregularly barred pattern
Albino—lack of pigment in the coat and eyes
Anemia—absence of red blood cells
Brown—pigmentation is chocolate in color
Curly—the fur and the whiskers are curved
Fawn—the coat color is tawny blue to fawn
Hairless—the rat loses its hair at about four weeks of age
Incisorless—the incisor teeth are missing
Kinky—the fur and the whiskers are kinky
Naked—the rat is hairless except for a very short fuzzy coat
Pink eye—the coat is yellow and the eyes pink
Red-eyed yellow—the coat is yellow and the eyes are red
Shaggy—the coat and the whiskers are curved
Silvered coat—a silver gray color
Stub—the tail is short and stubby
Waltzing—the rat runs in circles
Wobbly—the rat has an unsteady gait

Mating to Weaning Time and Other Vital Statistics

The figures below are averages for several species *in general*, and are meant only to provide a notion of relative time. Each species shows quite a range of variation. See under individual species for closer approximation.

Mating	⟶ Birth	⟶ Weaning
Mice	20 days	18 days
Rats	21 days	21 days
Rabbits	31 days	42 days

More specifically, the following table provides a relative notion of just where rats and their vital statistics stand in comparison with other small animals and *their* vital statistics:

	Gestation	Estrous cycle (days)	Age when breeding stops (months)	Number of babies produced by each female in 100 days
Mice	17-21	4-5	9	21-35
Rats	19-22	4-5	9	21-28
Rabbits	28-31	Rabbits are induced ovulators	24	30

Lifespan of Rats (Reported)

Cumming's giant rat	8 years 2 months 9 days
Muskrat	5 years 10 months
Emin's giant rat	5 years 6 months 11 days
Woodrat	4 years 8 months 15 days
Giant pouched rat	4 years 5 months 4 days
Bushy-tailed giant rat	4 years 3 months 28 days
Black rat	4 years
Cotton rat	2 years
Norway rat	2-3 years

In the wild, longevity does not always have a chance . . . less than five percent of Norway rats live more than one year in nature (luckily!).

Woodrats illustrate one of nature's balances: the woodrat has been observed to live for over three years in the wild, and that is quite long for wild rats. A lower reproduction rate, however, goes along with longer life. Conversely, shorter-lived rats have higher rates of reproduction.

SHIPMENT AND TRAVEL

Use gnaw-proof boxes (metal or wood reinforced with wire mesh) with perforated metal screened portholes on several or all sides. For trips over six hours, light wooden boxes are best, all of one side being of wire gauge or screen, and the other sides being reinforced with wire mesh. Lightweight metal boxes (with holes on all sides for ventilation) are fine for air shipment.

Wire shipping cages may be inadequate for these reasons: darkness is preferred by nocturnal species; privacy, too, is essential; wire offers no protection from inclement weather.

For rats which weigh thirty-five to fifty grams, provide six square inches in a box five inches high for each rat. This is for not more than twenty-five rats.

For rats which weigh fifty to one hundred fifty grams, provide eight square inches in a box five inches high for each rat. This, again, is not for more than twenty-five rats.

For adult rats, provide sixteen square inches per animal in a box five inches high and do not include more than twelve animals.

When receiving rats which have been shipped (hopefully and ideally no more than six to ten per box, although they may be shipped more densely packed than this), assume that they have been affected by changes in environment (psychological as well as climatic) and food. Overcrowding, cold drafts, noise and exposure to infections all may take a toll. Examine new animals at once. Isolate a new rat for about fourteen days before letting it associate with the stock which is already healthfully enjoying its home.

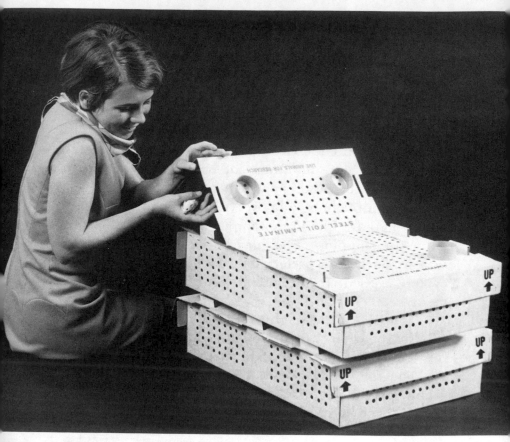

A high strength rat shipping container, designed and produced by Negus Animal Container Corp. of Madison, Wisc., utilizes .002 steel foil laminated inside the cardboard. This makes the carton dampness-proof, prevents chew-through and greatly increases the strength of the carton.

Before shipping rats, have the animals in good, well-fed condition. Place foods which have a high content of water in the shipping container. Potatoes, carrots, apples and lettuce are such foods.

Spread wood shavings or straw as litter in the shipping container. Arrange for the fastest carrier and avoid overcrowding the animals if they are shipped in hot weather.

RATS AS LABORATORY ANIMALS

Rats, along with mice and rabbits, constitute the major percentage of animals used in research; in the U.S.A., for example, mice accounted for about sixty five percent of vertebrates used in laboratories several years ago, rats about twenty-two percent, and rabbits about one point one percent. Laboratories use animals for research into the cause and cure of cancer and tumors, for testing the harmlessness (or dangers!) of food color or flavor additives, and for testing drugs intended for human consumption.

The thalidomide incident—when babies with limb deformities were born to mothers who had taken this drug during pregnancy—inspired a greater testing of new drugs in animals before the release of these drugs for use in human beings. Thalidomide was found to cause much the same teratogenic effect (abnormalities and malformations) in pregnant rabbits as it did in human beings. Rabbits, therefore, are being used quite extensively for this purpose, although rats are now beginning to contribute, too.

Large numbers of rats, rabbits, and mice, as well as other small animals, are used in the bioassay of literally thousands of prophylactic and therapeutic substances before the U.S. government will permit those substances to be released for sale to the public. Universities, research institutes and foundations, cancer research units, hospitals, U.S. Public Health laboratories, and the pharmaceutical industries are the organizations which use these animals for the above purposes.

Rats, rabbits and mice are also used for the laboratory diagnosis of diseases and of pregnancy in human beings.

And what about medical research on rats themselves and for themselves? Here is an experiment which helps manufacturers and breeders to raise healthier and happier rats. . . and which also contributes to the store of knowledge about nutrition in general. Such knowledge is often tapped for application in human medicine. This experiment helps to show that the genes from which we are made can attentuate as well as increase the effects of the environment around us (and the rats, too, of course). Two rats of equal age, sex and diet—a diet deficient in vitamin D—showed different results of the same diet. One rat grew fat and happy, but the other one developed rickets (underfed babies who do not receive ample vitamin D in their diet, may become bowlegged, a sign of rickets). The unaffected rat, the one which remained healthy, was from a strain of rats selected for (that is, bred for) resistance to rickets. The rat which developed rickets was from a strain of rats selected for susceptibility to rickets. It was the environmental stress—the deficient diet—which caused one to show its potential for the disease.

Rats and mice have served man well. . . not only as pets, but as laboratory animals where more than 30,000,000 give their lives every year. . . and as food. Sometimes fur from the larger rats is used to keep man warm. Photo by Harry V. Lacey.

OTHER ASPECTS OF THE RAT

Rats as Food

There are still aboriginal groups of non-technological people in our modern world who eat rats and rat-like creatures. The Chichimeca-Jonaz in the Mexican state of Guanajuanto, for example, are such a group. The Mexican census of 1950 reported that there were still five hundred and one of these Indians left. At that time, they ate rats along with maize (Indian corn), beans, chile and other foods. The Triquis, too, from the Mexican state of Oaxaca—a famous archeological site—have been observed eating rats.

In western civilization, however, periods of famine and deprivation have usually been necessary to bring forth some hardy (and hungry) souls who tried what is not normally food for them—rat meat. Rat meat brought a good price in 1798 in the French garrison at Malta. It was eaten at the siege of Paris in 1871. Also, it has been eaten by explorers (such as the Arctic explorer Kane whose note on rats appears earlier in this book). Prisoners of war, too, have found native rats a handy and tasty source of food. One authority on rodents who was once hoodwinked into eating rat disguised as terrapin, declared it to be delicious when told about the switch in menu afterwards . . . but said he recognized what he was eating because of the rat bones in his dinner plate. Large South American paca is a staple food in the jungle and is eaten by everyone.

The Ratters

The terriers, particularly the "rat terrier," have been bred as master killers of rats. They also excel at getting at and destroying other small, varmint-like animals. Terriers are even turned loose on foxes. The schnauzers, too, are well-known for their ratting ability. Terrier breeders still hold ratting trials—that is, rat-catching contests—for their dogs.

THE RAT IN FOLKLORE

The Pied Piper of Hamlen, of course, is well known as the fellow whose pipe-playing led the village children away and into the mountain because their parents would not pay his fee for having rid Hamlen of its rats (by piping them into the river).

Quite on the other side of the world, in Japan, the rat has the honor of having the first year of the oriental zodiac named after it. The animals which symbolize the years, days and hours are:

Rat
Ox or bull
Tiger
Rabbit or hare
Dragon
Snake or serpent
Horse
Sheep or goat
Monkey
Cock or bird
Dog
Wild boar

The Year of the Rat is called *Ne*—short for *nezumi*, which means rat. The rat in Japan has often been associated with the god of wealth—*Daikoku*—one of the seven gods of luck. Daikoku is portrayed sitting on two sacks of rice. . . with rats gnawing away at the bales. Some Japanese claim that this means that all wealth must be zealously guarded once it is acquired. Others say the only

Just look at this typical rodent. Does he look so dangerous that the fleas he carried killed thousands of people throughout the ages? Fortunately pet rats do not carry such dangerous fleas. . . nor do wild rats in most civilized countries.

connection is that Daikoku's festival happens to fall on the Day of the Rat. And still others believe that the rats are quite welcome to nibble a minute share of such wealth.

Rats, Plague and the "Pestilence"

First of all, every epidemic called "plague" or the "pestilence" in history was certainly not always what we understand today as plague—a disease caused by rat-borne infected fleas. Rodents and buboes in the descriptions of the Philistine plague which carried off 50,070 souls about 1070 B.C. indicate that this was the plague. 185,000 Assyrian troops died of the plague in 700 B.C. as Sancherib marched his army to battle the Egyptians. The

Justinian plague raged for fifty-two years (542 to 594), killing off fifty percent of the Byzantine Empire during the first twenty-three years of those fifty-two years of disease. The Black Death (1349 to 1351) took the lives of twenty-five percent of the European population.

Needless to say, long before the rats (and some other animals, depending on the locale) and their infected fleas were recognized officially as the cause (or carrier) of plague, many a "witch," "well-poisoner," religious or political dissenter, and innocent bystander was blamed for the disease and its ravages.

There must be disease among the rats, or rodents, themselves for human epidemics to develop. Dead and dying rats may or may not be evident during epidemics of plague among human beings. Or, if human beings live protected from the rat world, rats may die of disease without it passing on to man. "Rats and other subterranean creatures come up to the surface of the ground and behave as if they were drunk," wrote the Persian physician Avicenna, who lived from 980 to 1037.

Contemporary outbreaks of the plague are not unknown, even in the U.S.A. Before epidemiological reporting was done on a worldwide basis in conjunction with the World Health Organization, there was occasionally some hesitation in reporting such a devastating disease because of the consequent disinfection and quarantine imposed upon the whole area around a focus of infection. In India, for example, deaths were, in some cases, attributed to snakebite rather than to the plague, for that reason.

Albert Camus, a contemporary French writer (born in Algeria, North Africa) wrote a novel—*La Peste*—about plague in Oran. At first, no one recognized the disease. It was a strange disease, its victims being affected with violent bouts of fever, black patches on the skin, and running abscesses, or buboes. The physicians were hard put to realize that it could be the plague, a malady which was generally considered a thing of the far past.

Finally, the existence of the plague was recognized and announced by the authorities. Doors were locked. Guards were placed in ports and stations. Shipping was rerouted away from Oran. At last, the plague disappeared back into its obscure hideaway from which it had so silently emerged. Doors were reopened and public festivities were prescribed. In short, *La Peste* is a study of how a town reacts under something as frightful as a plague. It gives us, today, an insight into the confusion and turmoil of what must have reigned during plagues in bygone periods of history.

And, actually, localized outbreaks of plague do occur, or did occur until very recently, irregularly in Algeria, the scene of Camus' *La Peste*, particularly in port cities such as the one described by Camus. Countermeasures taken during an outbreak of the plague are anti-plague vaccination of exposed segments of the population, and stepped-up extermination of rats. The same species of flea (which plagues the rats it infects with the disease) is usually implicated in both plague *and* typhus outbreaks.

Giovanni Boccaccio's *Decameron* is another account (a somewhat ribald one) of what people do under stress of a public catastrophe—plague in Florence, Italy in the year 1348. In Boccaccio's story, a group of ten young men and women flee *Florence* in order to find refuge from the plague. They tell ten stories daily for ten days as they wait out the pestilence back in the city. And this communal effort at joviality is not too far removed from much more modern escapees from disaster; witness the community life engendered in bomb shelters and bomb-free zones of the Second World War, or the less sinister "hurricane parties" in the storm zones of the United States.

Rats are soft, furry, friendly animals that make great pets for the whole family. Especially for schoolkids who can stowaway their pet without the teacher seeing them!